The views expressed in this book are those of the experts who participated in two meetings on the subject of derived intervention levels, and do not necessarily represent the decisions or the stated policy of the World Health Organization.

DERIVED INTERVENTION LEVELS FOR RADIONUCLIDES IN FOOD

Guidelines for application after widespread radioactive contamination resulting from a major radiation accident

World Health Organization
Geneva
1988

ISBN 92 4 154233 0

The designations employed and the presentation of the material in this publication do not imply the expression of any opinion whatsoever on the part of the Secretariat of the World Health Organization concerning the legal status of any country, territory, city or area or of its authorities, or concerning the delimitation of its frontiers or boundaries. Where the designation ''country or area'' appears in the headings of tables, it covers countries, territories, cities, or areas.

The mention of specific companies or of certain manufacturers' products does not imply that they are endorsed or recommended by the World Health Organization in preference to others of a similar nature that are not mentioned. Errors and omissions excepted, the names of proprietary products are distinguished by initial capital letters.

Printed in Switzerland

88/7569—Phototypesetting—4500

CONTENTS

PREFACE

After the Chernobyl accident in April 1986 it was recognized that the available guidelines on the management of the consequences of a nuclear accident did not adequately cover the actions to be taken to protect the population in areas far removed from the accident site. Several meetings of international organizations were held to determine how this deficiency could be remedied. One outcome of the discussions was that WHO undertook to produce guideline values for derived intervention levels for radionuclides in environmental media, especially food, below which actions to reduce or avoid the potential health detriment would not be justified. Such guideline values could provide the basis upon which national authorities could implement their own derived intervention levels and so promote harmonization.

This book has been prepared from the reports of two 1987 WHO meetings on the subject of derived intervention levels. The report of the Working Group on Guideline Values for Derived Intervention Levels was widely circulated to national governments and comments were received from 24. A subsequent report was produced after the meeting of the Task Group on Guideline Values for Derived Intervention Levels. The participants at both meetings are listed in Annex 5, and their contribution to a particularly difficult topic is gratefully acknowledged. Special acknowledgement should be made of the contributions of Dr R. H. Clarke (National Radiological Protection Board, Didcot, Oxon, England) and of Dr D. Beninson (National Commission of Atomic Energy, Buenos Aires, Argentina). Dr Clarke performed considerable editorial and drafting duties, and Dr Beninson, as Chairman of both meetings, was largely responsible for encouraging the consensus that was eventually achieved in spite of the initially divergent views of participants.

WHO would also like to thank the Commission of the European Communities, the Food and Agriculture Organization of the United Nations, the International Atomic Energy Agency, and the Nuclear Energy Agency of the Organisation for Economic Co-operation and Development, which provided not only their expertise but also much of the data used in the calculation of derived intervention levels.

ABBREVIATIONS

CEC	Commission of the European Communities
FAO	Food and Agriculture Organization of the United Nations
IAEA	International Atomic Energy Agency
ICRP	International Commission on Radiological Protection
NEA	Nuclear Energy Agency of OECD
OECD	Organisation for Economic Co-operation and Development
UNSCEAR	United Nations Scientific Committee on the Effects of Atomic Radiation
USNRC	United States Nuclear Regulatory Commission
WHO	World Health Organization

Chapter 1

INTRODUCTION

Guidance on the management of the consequences of a nuclear accident is already available from a number of international organizations (e.g., IAEA 1985b, ICRP 1984b, WHO 1984). This guidance, however, concentrates on the actions required within a limited geographical area close to the accident site. After the major nuclear accident that occurred in Chernobyl, USSR, in April 1986, it became clear that additional guidance was needed for action over long time scales and for dealing with the widespread radioactive contamination that affected many countries at considerable distances from the accident site.

The actions undertaken by national authorities after the accident at Chernobyl *did* protect the public, but varied widely, even within similarly affected areas, and caused unnecessary confusion and unwarranted concern among members of the public (OECD/NEA 1987). Even within one country many organizations were necessarily involved, centrally, regionally, and locally, and the lack of coherent guidelines meant that approaches were sometimes inconsistent.

One of the major difficulties in areas away from the site of a nuclear accident concerns decisions on the safety of contaminated food and drinking-water. WHO, in close consultation with several other international organizations (see Preface), has therefore considered it appropriate to develop guidelines to assist national authorities in making decisions on the control of food in the event of widespread contamination by radionuclides resulting from a major nuclear accident. The guidelines are based on health protection principles and aim to minimize the risk to the population as a whole while taking due account of sensitive groups. They are intended to contribute to the national decision-making processes, and to provide a basis for a harmonized post-accident response by national authorities, while not imposing unnecessary constraints. It should be stressed that the guidelines relate only to the post-accident situation and are not applicable when a nuclear plant is operating normally. They concern only the "far field" (the area far removed from the accident site) and do not include guidance on the actions needed close to the accident site.

These guidelines deal solely with appropriate intervention levels for controlling exposure to radiation from contaminated food and drinking-water. However, it is important to remember that, after a nuclear accident, members of the public may be exposed to radiation externally and by inhalation of radionuclides as well as via ingestion of food and water. All routes of exposure should be taken into account by national authorities in the decision-making process.

Chapter 2

PRINCIPLES FOR THE CONTROL OF FOOD CONTAMINANTS

For the routine control of environmental pollution, limits are normally placed on the *release* of potentially harmful contaminants that may affect water, air, or soil and cause damage to the ecosystem or to human health. In addition, international guidelines have been developed defining acceptable *levels of intake* of contaminants — see for example the *Guidelines for drinking-water quality* (WHO 1984/5), and the reports of the FAO/WHO joint expert committees on food additives and pesticide residues.

In the case of foodstuffs, levels of contaminants in the product are usually regulated, but primary control is nevertheless aimed at the prevention of contamination.

Non-radioactive contaminants

Food control legislation and the principles that have evolved for its implementation over the past 80 to 100 years are designed both to reassure consumers of the quality, safety, and value of foods available on the market, and to provide food producers and the food-processing and marketing industries with basic criteria that, when met, facilitate trade in foods while safeguarding public health.

Certain contaminants such as mycotoxins, pathogenic microorganisms, heavy metals, and undesirable organic compounds occur in foods despite every effort to limit or prevent their occurrence. Some of the chemical contaminants are carcinogenic, and in such cases it is generally assumed that there will be a "no-effect level" on which to base an "acceptable level" for the contaminant in food. Toxicological information is usually available from the results of animal feeding studies, but it is rare to have data relating to health effects in human beings that can be used in the risk assessment of such contaminants. In these circumstances decisions on acceptable levels may include making a number of conservative assumptions, for example that all of a particular affected food item will be uniformly contaminated. A number of safety factors are also usually introduced so that the inherent uncertainties in the risk assessment procedure are allowed for.

In normal situations, food contamination can be controlled by, for example, limiting discharges from chemical industrial sites, controlling the manner in which chemicals such as pesticides or veterinary drugs are used, or minimizing levels of chemical toxins produced by fungal contamination of crops. A different situation exists after an accident. Although contamination of food from a chemical accident is usually fairly localized and affected foods can be readily controlled, it has always been recognized that, if the effects of condemning a source of contamination would be unduly severe, it might be necessary to accept less restrictive controls. It is rare for a chemical accident to have transboundary consequences, but this can occur (as after contamination of the Rhine in 1986). Generally the situation can be dealt with by bilateral agreements between national governments, so that contaminated food and water can be taken out of circulation or otherwise controlled fairly readily.

Radioactive contaminants

The principles for controlling radioactive contaminants in a normal situation, where efforts are mainly concerned with the prevention of contamination of the environment or foodstuffs, cannot be applied when an accident occurs. In the case of a nuclear accident causing widespread dispersion of radionuclides, controls have to be imposed after the event, with the objective of avoiding or minimizing the adverse effects. Nuclear accidents have special characteristics, in terms of the large areas and populations that may be affected. WHO has therefore considered it important to develop *public health guidelines* for the management of accidental radioactive contamination of food. These are based on risk assessment considerations that inevitably differ from those used to control human exposures in normal situations.

During the routine operation of a nuclear facility where the source of radiation is under control, radiation doses to members of the public are limited by applying restrictions to that source, for example by controlling the rate of release of radionuclides from the facility. These controls are generally based upon the ICRP system of dose limitation (ICRP 1977). During an accident, the source of exposure to radiation is by definition not under control and, as with chemical exposures, it is reasonable to accept that the normal system of dose limitation does not apply and that different criteria are necessary in decision-making.

Several studies have been made in human beings of the effects of acute and chronic high-dose exposure to radiation. Although the levels of exposure are not strictly comparable to those likely after a nuclear accident, these studies provide more accurate risk estimates than extrapolation from the animal feeding studies used as the basis of risk assessment for most chemicals. For planning radiation protection, the radiation risk estimates are generally applied without additional safety factors. However, as with chemicals, several pessimistic assumptions are built into the risk assessment; these introduce a margin of safety into the setting of intervention levels for radionuclides in food.

After a nuclear accident, as after a chemical accident, control measures may need to be less restrictive than in a normal situation if the total detriment arising from the accident and from the countermeasures is to be minimized. Since a major nuclear accident could result in contamination of areas of land thousands of kilometres from the source, international harmonization on control methods is needed to prevent discordant actions, which may cause unnecessary confusion.

The setting of acceptable levels of radionuclides in food will not itself ensure that exposures of individuals to radiation from food are kept below any agreed level. National authorities can determine this only by monitoring levels of contamination in the food and using actual food consumption data to assist in determining total intakes of the relevant contaminants.

Radiation accidents

The contamination resulting from a severe radiation accident creates problems for public health and other authorities that are very different in the vicinity of the accident and at a distance. The levels of exposure, the pathways of exposure, and the time frame over which the protective measures have to be implemented will depend on the specific circumstances of the accident, the season of the year, meteorological conditions, and agricultural practices.

Near field

In the immediate vicinity of the accident, radiation exposure rates can be very high, and in the worst case, can lead to acute health effects if protective measures are not implemented quickly. For convenience, the immediate area of the accident (with a radius of up to a few tens of kilometres) is referred to here as the "near field". If the accident is severe and involves the release of fission products as well as noble gases, then the pathways of exposure that predominate in the near field are whole-body external exposure from the airborne plume and from material deposited on the ground, and inhalation of radioactive material in the plume, which leads to radiation doses to internal organs. The external dose rate from radioactive material deposited on the ground may be dominant, especially if rainfall has enhanced the deposition; this deposited material, the transfer of its activity through food chains, and the inhalation of resuspended material then become important exposure pathways to be considered in decision-making. In the near field, urgent action such as sheltering, evacuation, and decontamination of individuals may be necessary. Administration of stable iodine tablets may also have to be part of the emergency response where radioiodines are released in significant quantities.

The risks, difficulties, and disruption that follow the implementation of various countermeasures after a nuclear accident differ widely, and thus the level of radiation dose at which a given countermeasure is introduced should be influenced by such considerations. Intervention levels must also be sufficiently flexible in application to be adapted to the particular conditions prevailing at the time of the accident.

In the guidance given to date by international organizations for the near field, an approach has been adopted with two tiers of reference dose levels for each protective measure. On radiation protection grounds, a lower level of dose has been recommended (IAEA 1985b, ICRP 1984b, WHO 1984) below which introduction of a particular countermeasure is not likely to be warranted, and an upper level of dose has been recommended above which implementation of the countermeasure should almost certainly be attempted. The upper level for the most difficult protective measure, evacuation, has been set at a whole body dose of 500 mSv if likely to be incurred in a short period of time. This corresponds to the level below which nonstochastic effects would not occur in a normal population. The lower intervention level of dose has been set at one-tenth of the upper level. For protective measures that are considered the easiest to be implemented (sheltering, control of foodstuffs), the upper and lower reference levels of dose to the whole body have been suggested at 50 and 5 mSv respectively. When preparing emergency plans, national authorities should set operational intervention levels between these upper and lower reference levels of dose.

Far field

The international guidance given so far does not specifically address the response to major nuclear accidents with substantial releases of radionuclides that spread to regions far from the accident site. Nevertheless, depending on the release characteristics and meteorological conditions, radionuclides can spread over very large areas — referred to here as the "far field". The problems thus created for the public health, agricultural, and other authorities differ from those in the near field.

The activity concentration of the plume lessens with time by radioactive decay of the radionuclides, by dilution, and by deposition on the ground, so that at large distances from the accident site (hundreds of kilometres) this source of potential contamination diminishes. There is relatively little deposition unless the passage of the plume coincides with rainfall. Local ground contamination is then very variable, but may be quite severe in some localities and contribute significantly to the external dose in the long term. However, outside these areas the plume contributes very little to the individual radiation dose either externally or internally through inhalation. In these circumstances, most of the exposure in the first years after the accident will occur from the incorporation of the deposited radionuclides into the human food chain. Thus, the doses in the far field tend to be less than in the near field and the principal mode of exposure changes from direct to indirect. Because the individual doses are lower at a distance than close to the accident, public health authorities will not need to introduce measures designed to reduce or avoid high-dose effects. Instead, food control and other appropriate authorities will need to consider effects on food availability and trade, while public health personnel will be more concerned with the potential future effects on the health of the exposed population.

Introduction of countermeasures

After a major nuclear accident, public health authorities may need to introduce measures to restrict the radiation doses received by members of the public so that the risks of adverse effects are small. Advice to limit exposure from ground deposition is not likely to be practicable except in very exceptional cases. On the other hand, measures may be needed to minimize the incorporation of radionuclides into foodstuffs produced in areas where local ground contamination is severe. Control over foodstuffs may have to be exercised for a very long time (months or even years) since radionuclides deposited on the ground may only slowly find their way into the food chain via animal feed or by being taken up by plants. Countermeasures can be taken either to minimize radioactive contamination in the food chain by, for example, feeding animals with the previous season's produce (at certain times of the year), or to prevent the consumption of contaminated food, for example by curtailing the sale of milk containing high concentrations of iodine-131 and converting it to cheese for later consumption. Such countermeasures may be taken as a precaution in the immediate aftermath of an accident, until a comprehensive assessment of the situation can be made.

Intervention levels in terms of *radiation dose* may be used as a basis for timely decision-making after an accident. In practice, however, the results of measurements made in the environment and on foods will be expressed in terms of *activity concentrations*, for example Bq/kg or Bq/l. "Derived intervention levels" (in terms of Bq/kg or Bq/1 rather than radiation dose) would therefore be more appropriate for authorities considering the introduction of particular protective measures. Such derived intervention levels can be determined (once intervention levels of dose have been set) from knowledge of physiological and metabolic processes in human beings, of the distribution of radionuclides in the body after different intakes, and of the resulting radiation doses to various body organs. Necessarily, the levels are based on "reference persons" of different age groups and on average intakes of air, food, and water. Because derived intervention levels relate to concentrations of radionuclides in the different environmental media, such as air, water, land, and food, and because such concentrations can be readily measured, swift action can be taken to minimize the exposure of individuals to radiation should the derived intervention levels be exceeded. In the far field without ground deposition of radionuclides, the largest contribution to the radiation doses received by individuals is through the food chain. Accordingly, the derived intervention levels considered here relate only to foods consumed by the population.

Chapter 3

PRINCIPLES FOR THE CONTROL OF RADIATION DOSE

ICRP (1977) recognizes two quite different conditions of exposure to radiation:

(*a*) when the exposure is foreseen and can be limited by control of the source and by the application of the system of dose limitation, including the development of satisfactory operating procedures; and

(*b*) when the source of exposure is not subject to control, so that any subsequent exposure can be limited in amount, if at all, only by remedial actions.

The second condition is the relevant one after accidents that have released radioactive materials to the environment. However, it is useful to review the principles of control of radiation dose in normal situations so that the differences in decision-making after an accident can be clearly distinguished.

Normal operations of nuclear facilities

ICRP has recommended that radiation doses due to discharges of radioactive material to the environment (resulting from justified use of radioactive materials) be as low as reasonably achievable. In any case, no individual should be exposed above the appropriate dose limit. In this situation the source (discharge) is controlled and most national authorities ensure that a single source does not give more than a small fraction of the relevant dose limit to the most exposed "critical group". As a result, doses to the most exposed individual members of the public are low, and at present, population doses are also low. In 1980, global averages for individual radiation doses from the nuclear power cycle were about 0.1 μSv per year, which is less than 0.01% of those due to natural background radiation (UNSCEAR 1982). The requirement to keep doses as low as reasonably achievable has therefore resulted in mean doses to the population that are much lower than the dose limits. From the "fatal cancer risk per unit dose" figure suggested by ICRP (1987) of about 2×10^{-2} Sv^{-1}, the average presumed risk of fatal cancer for an individual can be calculated at less than 10^{-7} per year from controlled sources of nuclear power production.

After an accident

Although the system of dose limitation does not apply to situations where the exposures cannot be limited by control of the source, its underlying principles can be used to formulate a rationale for intervention planning. This rationale was summarized by ICRP (1984b) as follows:

"(*a*) Serious non-stochastic effects should be avoided by the introduction of countermeasures to limit individual dose to levels below the thresholds for these effects.

(*b*) The risk of stochastic effects should be limited by introducing countermeasures which achieve a positive net benefit to the individuals involved. This can be accomplished by comparing the reduction in individual dose, and therefore individual risk, that would follow the introduction of a countermeasure with the increase in individual risk resulting from the introduction of that countermeasure.

(*c*) The overall incidence of stochastic effects should be limited, as far as reasonably achievable, by reducing the collective dose equivalent. This source-related assessment may be carried out by cost–benefit analysis techniques and would be similar to a process of optimization in that the cost of further reducing the health detriment in the affected population is balanced against the cost of further countermeasures."

The three principles apply throughout the progression of a large nuclear accident, but their relative importance changes as the potential for large radiation doses disappears and the individual risk becomes progressively smaller. In the long term and far field the decision whether to introduce a countermeasure, or to continue one already established, should be based on principles (*b*) and (*c*).

The dose limits recommended for controlled situations are not intended to apply in post-accident situations. Several questions have been raised, however, about the application of dose limits, after an accident, to the distribution of foodstuffs in the far field and long term. The main issue raised is essentially that, if contaminated foodstuffs represent the source of exposure to radiation, the situation is controllable by banning distribution of such foodstuffs, and therefore the full system of dose limitation should apply, including the dose limits. This is, however, a misinterpretation of the intent of the ICRP recommendations (ICRP 1977). The dose limits recommended by ICRP were meant to apply to the combination of sources stipulated and to the control of those sources, particularly in the design of nuclear installations and planning of operations. They were not intended to apply to radionuclides present in the environment due to an accident.

The situation may be compared with that of exposure from natural radiation sources, which are not included in the dose limits. ICRP (1984a) recognized that almost all exposures to natural sources of radiation are controllable to some extent, but the degree of controllability varies very widely, as do the complexity, cost, and inconvenience of the possible control measures. *Controllability* must therefore be a major factor in any system of intervention. From this point of view, there is a clear difference between *existing* exposure situations, where any action would have to be remedial, and *potential* exposures in the future, which can be subject to limitation and control at the stages of decision and planning.

The same applies to artificial radionuclides in the environment. Design and planning can control and limit radioactive effluent releases (and therefore *potential* exposures) at the source, but in *existing* situations exposures can be altered only by taking some remedial action in the environment. As an aid to deciding whether such remedial action should be initiated, the ICRP recommends the use of an intervention level specific to the action being considered. An intervention level is not determined by limits intended for future situations nor, specifically, by the primary dose limits recommended by the ICRP for members of the public.

Radiation dose in successive years and from more than one accident

The individual radiation dose in the second and subsequent years after a nuclear accident is likely to be considerably less than that received in the first year. In this case it is not necessary to revise the intervention dose levels or derived intervention levels for successive years, since the situation in individual countries will vary. In the rare situation in which the annual radiation dose has not fallen significantly by the second or third year, it is for the competent authorities to consider the specific circumstances and decide whether additional action is needed.

The possibility of two nuclear accidents, each resulting in significant contamination of food in the same area and time period, is remote. However, if such a situation should occur, radiation exposures from both accidents should be considered together.

Chapter 4

JUSTIFICATION OF THE DOSE LEVEL FOR INTERVENTION

The individual dose criterion

ICRP has suggested that measures to restrict the distribution of foodstuffs should be considered if the dose to the individuals who would be affected by the countermeasure is projected to exceed 5 mSv effective dose equivalent as a result of intakes in the first year after an accident (ICRP 1984b, Appendix C). Although this figure was primarily intended to apply in the near field, it is also appropriate in the far field.

The risk figure for fatal cancers suggested by ICRP (1987) is about 2×10^{-2} Sv^{-1} averaged over age and sex. An average individual exposed to 5 mSv as a result of ingestion of radioactively contaminated foodstuffs in the first year after a radiation accident therefore has a notional lifetime risk of 1 in 10 000 (10^{-4}). This level of risk is some three orders of magnitude greater than the average individual risk of fatal cancer resulting from routine operations of nuclear power establishments. It may be compared with the level of risk from another widespread environmental hazard — radon in houses. In this case no international agency or individual country has proposed a level of dose at which remedial measures should be taken against the *existing situation* that is as low as 5 mSv per year. ICRP (1984a) has recommended remedial measures that are simple and non-intrusive at dose levels from radon exceeding 20 mSv per year, although for the planning of measures to limit future exposures a lower level (10 mSv per year) is proposed. A WHO expert group recommended that simple remedial actions to reduce concentrations of radon should be considered if the annual effective dose equivalent is more than 8 mSv, and that at doses of 32 mSv per year remedial actions should be taken without long delay (WHO 1986).

An intervention level of individual dose of 5 mSv can also be justified on a comparison with natural background radiation and its variation. UNSCEAR (1982) gave an average figure of 2 mSv per year for normal exposure to all natural background sources for adults and 3 mSv per year for children. About half of this average effective dose equivalent is due to radon daughter exposure of the lung and that component varies over three orders of magnitude for many developed countries. A better comparison is with the variation in other components of natural background radiation that expose the whole body; these produce total radiation doses between 1 and 10 mSv per year (UNSCEAR 1982). An intervention level of dose of 5 mSv is therefore comparable with the global variation in the effective dose equivalent due to natural radiation sources.

If an intervention level of 5 mSv is set for the effective dose equivalent, a thyroid dose of 167 mSv is implied when this organ alone is irradiated. This dose is considered too high, given the incidence of nonfatal cancer following thyroid irradiation and the potential of iodine-131 for irradiation of the thyroid alone. A limitation of the thyroid dose to 50 mSv would make allowance for these factors.

After the Chernobyl accident, the average levels of radioactive material in individuals were substantially lower than was predicted from measurements of the amount of radioactive material deposited where those individuals lived (Fry & Britcher 1987, Meekings 1987). This was due to the complexity of the food web, rather than the failure of models to predict the radionuclide content of food produced in the contaminated area. Because most individuals obtain components of their diet from wide areas, only a fraction of the food consumed will be contaminated at the level corresponding to the deposition level where they live. It can be concluded that, if an intervention level of dose of 5 mSv is applied, mean doses to individuals are likely to be significantly lower than 5 mSv.

One additional point, concerning health risk, should be considered. ICRP (1986) has pointed out that serious mental retardation could result from exposure of the fetus during the period 8–15 weeks after fertilization, with a risk of 0.4 Sv^{-1}. If exposure was uniform over a year, 5 mSv would incur a risk of serious mental retardation of the order of 3×10^{-4} in a child who was exposed as a fetus. However, there is evidence that a significant threshold may exist for this effect (ICRP 1987), and if this is so, no extra precautions need to be introduced, because the threshold may be as high as several hundred millisieverts. Until the presence or absence of a threshold is confirmed, national authorities may wish to consider mental retardation as a possible stochastic effect that confers critical-group status on the fetus at the stage of development 8–15 weeks after fertilization.

The population dose criterion

As discussed in the previous section, the use of an intervention level of dose of 5 mSv in a year will ensure that individual risks are adequately controlled. However, reliance solely on an individual protection criterion could be criticized on the grounds that it makes no allowance for the number of people exposed and for the overall detriment to society. This question is particularly important in the far field where individual doses, and therefore individual risks, are low. The total societal detriment needs to be assessed in order to determine whether intervention at a lower level of individual dose than 5 mSv is justified to reduce this detriment.

The total societal detriment is dependent on the number of persons exposed and thus the collective dose. Restriction of this detriment is the objective of the basic radiological protection principle of keeping doses "as low as reasonably achievable" — that is, of reducing the societal detriment to a level such that further reductions would not be warranted. The decision as to what is "as low as reasonably achievable" is based on a balancing of the cost of the intervention and the cost of the detriment saved. The "individual" reference level of dose (of 5 mSv) constrains the cost-benefit analysis since this reference level must not be exceeded. However, where the cost of the intervention is low, it may well be justified to intervene to reduce the detriment at a lower level of dose. Nevertheless, only countermeasures that produce clearcut reductions in dose are warranted since there are always uncertainties in cost-benefit analyses.

The cost-benefit analysis involves assigning monetary costs both to the radiation detriment and to the imposed countermeasure, and calculating an intervention level of dose that will minimize the sum of the costs, i.e., that will minimize the total negative impact on society. The calculated intervention level should be accepted *only* if it lies below the individual reference level; if it lies above, it should be rejected. For intervention at an annual level of dose H, the point of balance at which further limitation to the health detriment of a population is not warranted on radiological protection grounds is given by:

$$H = \frac{C}{\alpha}$$

where C is the cost of maintaining the countermeasure per person and per unit time and α is the cost assigned to the unit of collective dose (see Annex 1).

The value of α can be regarded as the monetary sum that society is willing to expend to avoid the harm caused by a unit collective effective dose equivalent. To facilitate the comparison of options to control health detriment, the IAEA has recommended that, for exposures resulting from the transboundary movement of radioactive materials, α should not be less than US$3000 per manSv (IEAE 1985a). In making this recommendation, IAEA recognized that national authorities in countries with a high per caput gross national product or with high goals for radiation safety may wish to use a larger value of α. Indeed, values of α ranging from US$2000 to US$20 000 per manSv have been used in radiation protection optimizations in different countries.

The parameter C reflects the economic and social cost of taking action to protect members of the public. At its minimum, C (the cost per person and unit time) is equal to the product of the cost of a foodstuff per unit mass ($ kg^{-1}) and the average individual consumption rate (kg per year). To this value, the costs of social or trade disruption and the administrative costs of implementing the countermeasure should be added. The resulting cost is likely to be in the range of US$20–200 per person per year depending on the value of the commodity, the cost of social disruption, and the cost of administration. This range is illustrated well by foodstuffs with production costs of less than a dollar per kilogram and with consumption rates of the order of 100 kg per year. If a control measure affects international trade, a total cost should be calculated that includes the hidden and global costs of taking the action; the cost assessment should not be restricted to costs incurred nationally or locally within a country.

If C is US$20 per person per year, then the annual dose below which countermeasures are unjustified is 10 mSv for α = US$2000, which is at the bottom end of the range of values of α. If the social and economic disruption is higher, and one takes account of other costs such as global trade factors, C may be US$200 per person per year. This also gives an intervention level of 10 mSv if a high value of α of US$20 000 is assumed, as is reasonable for more developed countries. Indeed, since the calculated intervention level (H) depends on the ratio of the economic and social cost of further disruption (C) to the amount society is prepared to pay to avoid health detriment (α), and since both these parameters tend to move in the same direction, H is relatively insensitive to changes in C and α.

It may be concluded that application of the population dose criterion will result in intervention levels of dose in the region of several millisieverts per year. This supports the choice of 5 mSv as the intervention level on the basis of the individual dose criterion. The derived intervention levels listed in Chapter 6 have therefore been calculated on the basis of an intervention level of 5 mSv in a year.

Chapter 5

INPUT DATA FOR DERIVED INTERVENTION LEVELS

The purpose of this chapter is to summarize the data on food consumption and on relevant radionuclides that are needed for the calculation of derived intervention levels.

Food consumption data

The value of the derived intervention level (*DIL*) for a food varies inversely with the mass of food consumed and the dose per unit

FIG. 1

PLOT OF FOOD CONTAMINATION VERSUS FOOD CONSUMPTION FOR ^{137}Cs TO GIVE A DOSE OF 5 mSv ASSUMING A DOSE CONVERSION FACTOR OF 1.3×10^{-8} Sv/Bq

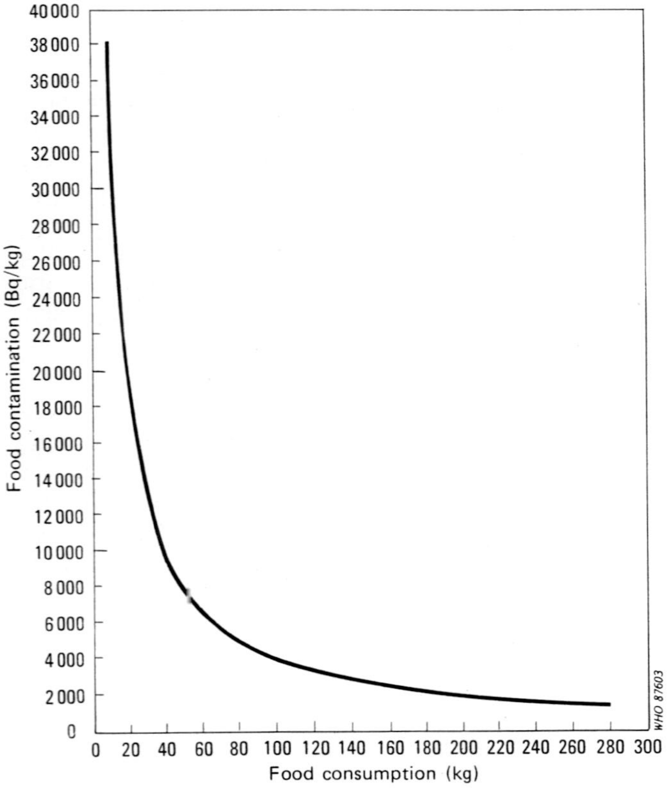

intake factor, and directly with the reference level of dose used, so that

$$DIL = \frac{RLD}{md}$$

where RLD = Reference (intervention) level of dose (Sv/year)
m = Mass of food consumed annually (kg/year)
d = Dose per unit intake (Sv/Bq).

Figure 1 shows how, for caesium-137, the intake of contaminated food that gives a dose of 5 mSv to an adult varies with the level of contamination. For low intakes ($<$ 20 kg) high contamination is required before the reference level of dose is exceeded. For low-consumption food items the contribution to total radiation dose is therefore small and guideline values for derived intervention levels will be extremely high.

Food consumption data have been reviewed for about 140 countries and areas. The averages for food groups that are usually consumed in quantities greater than 20 kg per year are tabulated in Annex 2. These food groups are: cereals, roots and tubers, vegetables, fruit, meat, fish (in certain parts of the world), and milk. On the basis of patterns of food consumption, geographical location, and existing FAO groupings, the countries and areas have been divided into eight diet types: African, Central American, Chinese, Eastern Mediterranean, European, Far Eastern, North African, and South American. Averages have been computed for each diet type and for each food group in kilograms per year (Table 1). There is quite a wide spectrum of consumption values for certain food items (e.g., milk) but for others (e.g., cereals), the variation is much smaller. Within countries, individual consumption patterns also vary.

Table 1. Food consumption patterns for different diet types (per caput in kilograms per year)

Diet type	Cereals	Roots and tubers	Vegetables	Fruit	Meat	Fish	Milk
African	127.4	134.8	25.9	45.2	16.8	15.0	28.9
Central American	113.2	46.0	38.8	98.6	42.3	18.7	82.2
Chinese	171.7	85.8	85.2	5.5	15.0	8.9	1.9
Eastern Mediterranean	188.2	19.3	91.6	101.5	30.4	8.4	74.3
European	121.1	72.7	86.7	81.4	75.3	20.2	154.9
Far Eastern	206.6	28.4	54.3	48.3	21.5	24.4	33.6
North African	161.9	20.0	63.4	63.9	24.0	7.3	77.2
South American	129.5	67.6	34.2	83.2	48.4	14.4	70.7

As the consumption rates vary widely between the different regional diet types, a method is required to establish representative intake rates for the purpose of calculating derived intervention levels. Global average consumption rates do not adequately reflect the maximum intake rates. However, using the maximum intake rate for each foodstuff would give too high a total consumption rate for calculating the individual dose. The usual practice in radiation protection is to use data from a reference group whose habits are representative of the most exposed individuals. The approach proposed here is to normalize the maximum regional intake rates for each food category to a representative total food consumption rate. The FAO figure for total food consumption of 550 kg per year (not including water) has been adopted for this purpose.

Table 2. Regional maximum food consumption and normalized rates (per caput in kilograms per year)

	Cereals	Roots and tubers	Vegetables	Fruit	Meat	Fish	Milk
Regional maximum[a]	205	135	90	100	75	25	155
Normalized	140	100	60	70	50	15	105

[a] Figures have been rounded off for convenience.

The normalized values given in Table 2 reflect food consumption by a hypothetical reference group and do not represent any single individual's actual consumption. The normalized values compensate for the variation in food consumption, so that maximal regional intakes are better represented than they would be if global averages were used. Contaminated fish is unlikely to be important as a source of radiation exposure in most regions. However, in areas where the consumption of freshwater fish is high, fish as a food component can assume importance as a source of exposure, and is therefore included even though its normalized consumption rate does not exceed 20 kg per year.

A value of 700 l per year was adopted as representative of drinking-water consumption for adults.

Relevant radionuclides

The range of radionuclides that may be released into the environment as the result of a nuclear accident will depend upon the type of plant involved and the nature and severity of the accident. The range may include long-lived and short-lived fission products, actinides, and activation products. In principle, it is possible to establish derived intervention levels for all radionuclides; in practice, however, for the far field the calculation can be limited to those of the greatest potential significance in the contamination of the food chain and environmental media after an accidental release.

The important radionuclides likely to be released by any particular installation can be identified from a safety analysis of the plant, and estimates made of the composition and relative proportions of radionuclides that would be released in different accident situations. The range and amount of radionuclides released may differ markedly between different types of installations (reactors, reprocessing plants, fuel fabrication plants, isotope production units, etc.) and should, therefore, be determined in advance for the particular facility and the exposure pathways of concern.

The guidance provided in this book is concerned with implementing protective measures that are appropriate in the far field, at distances significantly beyond the area of immediate impact of an accident. Emphasis is therefore placed on the radionuclides most likely to be encountered as a result of reactor or reprocessing plant accidents and on those whose longer half-lives could make them of radiological concern in the far field should they enter the food chain.

A number of studies have been made to identify the pathways for accidental releases from installations in the nuclear fuel cycle and the radionuclides of potential radiological significance (Alpert et al. 1986, Charles et al. 1983, USNRC 1975). Table 3 lists the radionuclides most likely to be of significance in food pathways after an accident at a reactor or reprocessing plant. These are the radionuclides for which derived intervention levels would be appropriate.

Table 3. Radionuclides that may be of significance in food pathways after a nuclear accident

Strontium-90
Iodine-131
Caesium-134
Caesium-137
Plutonium-239

Dose per unit intake factor[1]

The radiation dose to an individual from the intake of a radionuclide depends on the activity ingested and on other variables such as age, metabolic kinetics, and size, for example the dose to an infant will usually differ from that to an adult for the same activity ingested. Where the dose to a critical age group in the population is of interest, differing dose per unit intake factors have been developed for the different age groups. The recommended committed effective or organ (thyroid) dose equivalents per unit intake (Sv/Bq) for three age groups (one-year-olds, 10-year-olds, and adults) are shown in Table 4. The dose to a critical group should be calculated from the age-specific dose factor and the food consumption data for that group.

The doses per unit intake recommended here in Table 4 may differ from those recommended by ICRP (1979), which are for adults exposed to radiation at work and are based on "reference man"; the ICRP values cannot be applied indiscriminately to members of the public where the age distribution is less uniform. This point is discussed more fully in Annex 3.

Table 4. Age-specific dose per unit intake factors for different radionuclides (Sv/Bq)[a]

Radionuclide	One-year-olds	10-year-olds	Adults
Strontium-90[b]	1.1×10^{-7}	4.0×10^{-8}	3.6×10^{-8}
Iodine-131[c]	3.6×10^{-6}	1.0×10^{-6}	4.4×10^{-7}
Caesium-134[b]	1.2×10^{-8}	1.2×10^{-8}	2.0×10^{-8}
Caesium-137[b]	1.0×10^{-8}	1.0×10^{-8}	1.3×10^{-8}
Plutonium-239[b]	2.4×10^{-6}	1.4×10^{-6}	1.3×10^{-6}

[a] Interim values adopted until values have been recommended by ICRP.
[b] Committed effective dose equivalent.
[c] Committed dose equivalent to the thyroid.

The radionuclides of interest can be divided into two broad groups on the basis of dose per unit intake factors: Class I, with a high effective dose per unit intake factor of the order of 10^{-6} Sv/Bq (e.g., plutonium-239 and other actinides), and Class II, with a low effective dose per unit intake factor of the order of 10^{-8} Sv/Bq, which will include all other radionuclides of interest. These values are used with the normalized food consumption data discussed earlier in the chapter to calculate derived intervention levels.

[1] Also known as the "dose conversion factor".

Chapter 6

GUIDELINE VALUES

Guideline values for derived intervention levels for radionuclides in food have been calculated from the intervention levels of dose and input data discussed in Chapters 4 and 5, i.e., on the basis of:

(*a*) a maximum annual committed effective dose equivalent of 5 mSv (but a maximum committed dose equivalent of 50 mSv for the thyroid);

(*b*) a normalized hypothetical diet based on representative values of higher-than-average consumptions of different food components in different parts of the world; and

(*c*) two dosimetric classes of radionuclides, one for actinides and the second for other nuclides.

As the dose to an infant from a particular radionuclide may be significantly higher than the dose to an adult, separate guideline values have been developed for the infant on a milk diet.

The guideline values for derived intervention levels for seven food groups and for drinking-water are presented in Table 5. The calculated values are of necessity based on an effective dose of 5 mSv for each nuclide *in isolation in a single food category*, since it is not possible to generalize regarding which nuclides will be most important in each food category after an accident. It is recommended that the guideline values be modified as appropriate to the particular accident situation. A number of examples are developed in Annex 4 to demonstrate how derived intervention levels can be calculated to take account of the summed effects of different nuclides in different food categories.

Table 5. Guideline values for derived intervention levels (Bq/kg)

Class of radionuclide	Cereals	Roots and tubers	Vegetables	Fruit	Meat	Milk	Fish	Drinking-water
I: High dose per unit intake factor (10^{-6} Sv/Bq)	35	50	80	70	100	45	350	7
II: Low dose per unit intake factor (10^{-8} Sv/Bq)	3500	5000	8000	7000	10 000	4500	35 000	700

Guideline values for infants are presented in Table 6. These were developed as it was felt that the guideline values based on the rounded-off dose per unit intake factors (10^{-6} and 10^{-8} Sv/Bq) and the normalized hypothetical diet did not adequately protect this group. The values are based on an assumed annual consumption of 275 kg milk and 275 kg water, and on the dose per unit intake factors applicable to infants.

Table 6. Guideline values for derived intervention levels for infants (Bq/l)

Strontium-90	160
Iodine-131	1600[a]
Caesium-137	1800
Plutonium-239	7

[a] Based on a mean life of 11.5 days for iodine-131 and an organ dose of 50 mSv to the thyroid.

The values set for drinking-water for general application (Table 5) adequately protect the infant, except from strontium-90, for which a separate value is required (160 Bq/l).

The guideline values in Tables 5 and 6 have been rounded off for convenience. It is considered that, since values have been recommended for both the general population and infants, children will be sufficiently protected.

If several radionuclides have been released during an accident and several food components are contaminated, then national authorities should allow for the adding of dose contributions. It is not possible to generalize on the relative importance of different foodstuffs or nuclides because these depend on the accident and the location of the deposition, so no specific values have been set. National authorities may produce their own "specific" derived intervention levels by means of a formula that ensures that the total reference level of radiation dose (5 mSv) is not exceeded when a number of foodstuffs are contaminated by one or more radionuclides, that is,

$$\sum_{i} \sum_{f} \frac{C(i, f)}{DIL(i, f)} \leqq 1$$

where $C(i,f)$ is the activity concentration of isotope i in foodstuff f and $DIL(i,f)$ is the derived intervention level of isotope i in foodstuff *f*. Some examples of the use of the formula are given in Annex 4.

The guideline values apply at the point of consumption and to the form of the food as it is consumed. Other derived intervention levels need to be calculated if control measures are to be instituted at a different stage of the food production or distribution. The consumption of minor foodstuffs such as spices, herbs, and tea is low and, if they were contaminated, vast quantities would need to be consumed before a significant contribution to the individual dose of 5 mSv was approached (see Fig.1, page 21). Accordingly, for such low-consumption food items, an increase in the derived intervention level for roots and tubers by a factor of 10 would offer conservative protection, i.e., 500 Bq/kg for Class I radionuclides and 50 000 Bq/kg for Class II radionuclides.

REFERENCES

ALPERT, D.J. ET AL. *Relative importance of individual elements to reactor accident consequences assuming equal release fractions.* Albuquerque, NM, Sandia National Laboratories, 1986 (Report NUREG/CR-4467 (SAND85-2575)).

CHARLES, D. ET AL. *Contributions of nuclides and exposure pathways to the radiological consequences of degraded core accidents postulated for the Sizewell PWR.* London, HMSO, 1983 (Report NRPB-M100).

FRY, F.A. & BRITCHER, A. Doses from Chernobyl radiocaesium. *Lancet,* **2**: 160–161 (1987).

IAEA. *Assigning a value to transboundary radiation exposure.* Vienna, IAEA, 1985a (Safety Series No. 67).

IAEA. *Principles for establishing intervention levels for the protection of the public in the event of a nuclear accident or radiological emergency.* Vienna, IAEA, 1985b (Safety Series No. 72).

ICRP. Recommendations of the International Commission on Radiological Protection. *Annals of the ICRP,* **1**(3) (1977) (ICRP Publication 26).

ICRP. Limits for intakes of radionuclides by workers. *Annals of the ICRP,* **2**(3/4) (1979) (ICRP Publication 30, Part 1).

ICRP. Principles for limiting exposure of the public to natural sources of radiation. *Annals of the ICRP,* **14**(1) (1984a) (ICRP Publication 39).

ICRP. Protection of the public in the event of major radiation accidents: principles for planning. *Annals of the ICRP,* **14**(2) (1984b) (ICRP Publication 40).

ICRP. Developmental effects of irradiation on the brain of the embryo and fetus. *Annals of the ICRP,* **16**(4) (1986) (ICRP Publication 49).

ICRP. Statement from the 1987 Como meeting of the International Commission on Radiological Protection. *Radiological protection bulletin,* **86**(suppl.) (1987).

MEEKINGS, G.F. Radionuclide intakes in food: comparison between calculated and measured values. In: *Proceedings of the CEC International Seminar on Foodstuffs Intervention Levels following a Nuclear Accident, Luxembourg, 27–30 April 1987.* Luxembourg, CEC, 1987, pp. 251–253 (unpublished preprint EUR 11232).

OECD/NEA. *The radiological impact of the Chernobyl accident in OECD countries.* Paris, OECD/NEA, 1987.

UNSCEAR. *Ionising radiation: sources and biological effects. United Nations Scientific Committee on the Effects of Atomic Radiation 1982 Report to the General Assembly.* New York, United Nations, 1982.

USNRC. *Reactor safety study: an assessment of accident risks in US commercial nuclear power plants.* Washington, DC, USNRC, 1975 (Report WASH-1400, App. VI (NUREG-75/014)).

WHO. *Nuclear power: accidental releases — principles of public health action.* Copenhagen, WHO, 1984.

WHO. *Guidelines for drinking-water quality. Vol.1. Recommendations. Vol.2. Health criteria and other supporting information. Vol.3. Drinking-water quality control in small-community supplies.* Geneva, WHO, 1984/5.

WHO. *Indoor air quality: radon and formaldehyde. Report on a WHO meeting.* Copenhagen, WHO, 1986 (Environmental Health Series No. 13).

Annex 1

THE POPULATION DOSE CRITERION

The decision to introduce a countermeasure should be based on a balance of the radiation risk avoided and the risks and disadvantages caused by the countermeasure itself. The intervention level determines the initiation of a particular type of remedial action and it is implicit in its selection that the exposed individual should be put in a "better" position by the remedial action, i.e., that the health risk to the individual should be reduced at a reasonable cost in financial and social terms.

The decision to introduce a remedial action can be based on a conceptual cost–benefit analysis, such that the action is taken only if the net benefit is positive. This benefit could be expressed as

$$B = Y_{o} - Y_{I} - R - X$$

where

B is the net benefit,

Y_{o} is the radiation detriment cost if the remedial action is *not* taken,

Y_{I} is the remaining radiation detriment cost if the remedial action is carried out,

R is the detriment cost caused by risks due to the remedial action itself, and

X is the cost of the remedial action.

In practice it is difficult to quantify all the terms in the equation and it is necessary to make subjective value judgements, similar to those involved in most social and economic decisions.

If intervention is decided on, then the selection of the appropriate intervention level may maximize the net social benefit. If the detriment cost due to the risks from the countermeasure itself is independent of the intervention level, the optimization condition is

$$\frac{dX}{dI} + \frac{dY_{I}}{dI} = 0$$

where I is the intervention level or any derived intervention level. The cost of the countermeasure (X) may appear minimal if assessed only locally; ideally, however, the cost assessment should be conducted on a wide basis and should include the costs of production of the foodstuff wherever they are incurred, as well as those of transportation and administration.

If one assumes that in the far field and long term only the so-called α-term of the detriment cost is relevant for optimization, the optimizing condition can be expressed as

$$\frac{\mathrm{d}X(I)}{\mathrm{d}I} + \alpha \frac{\mathrm{d}S_I}{\mathrm{d}I} = 0$$

where $X(I)$ is the part of the countermeasure cost that is a function of the intervention level, α is the monetary value assigned per unit collective dose, and S_I is the collective dose remaining after the countermeasure has been applied.

A simple optimization procedure can be formulated for situations where the intervention is fully effective while it is applied, and where both the collective dose and the cost of the countermeasure are proportional to the number of individuals affected by the countermeasure. If the countermeasure is applied for a time τ, during which individual doses are zero, and then removed, the cost and the remaining collective dose can be expressed as

$$X = CN\tau$$

$$S_I = N \int_{\tau}^{\infty} H(t)\,\mathrm{d}t$$

where C is the cost of the countermeasure per person and per unit time, $H(t)$ is the individual dose as a function of time if the countermeasure is not applied, and N is the number of people affected by the countermeasure.

The optimizing condition described previously can now be expressed as

$$\frac{\mathrm{d}X(\tau)}{\mathrm{d}\tau} + \alpha \frac{\mathrm{d}S_I}{\mathrm{d}\tau} = 0$$

and therefore

$$H_o(\tau) = \frac{C}{\alpha}$$

where $H_o(\tau)$ is the optimum value for the individual dose intervention level. It should be noted that the ratio C/α is expected to be more insensitive to geographical location than either C or α, because richer countries where C would be higher are likely to assign higher values to α.

Annex 2

FOOD CONSUMPTION DATA

Food consumption data have been reviewed for about 140 countries and areas, and data for food components with a consumption rate of more than 20 kg per year (as well as for fish, see page 23) have been tabulated. Countries and areas with similar food consumption patterns have been grouped into regional types, from which an average consumption has been derived (Tables 7–14). Eight average dietary patterns can be discerned and these are listed in Table 1, page 22.

The following abbreviations are used in the tables:

DS	Diet Survey
FBS	Food Balance Sheet
HE	Household Expenditure
INT	Interview
INV	Inventory
RC	Recall
WS	Weighing Survey.

Dashes in table columns indicate that no data are available.

Table 7a. African-type cereal-based diet (per caput in g/day)

Country or area	Type of data	Cereal	Roots and tubers	Vegetables	Fruit	Meat	Fish	Milk	Reference
Botswana	FBS	404.8	27.7	76.7	48.6	90.0	4.2	309.1	1
Burkina Faso	FBS	441.3	41.2	24.9	20.5	26.9	4.2	59.6	1
Cape Verde	FBS	528.9	146.6	43.9	105.8	19.3	65.8	162.4	1
Gambia	FBS	529.5	27.3	33.8	14.3	39.3	65.2	62.3	1
Kenya	FBS	374.5	185.7	60.2	72.7	55.6	8.7	161.8	1
Kenya	HE	422.0	219.0	60.0	81.0	37.0	7.0	131.0	2
Lesotho	FBS	611.6	16.6	63.3	44.5	54.0	8.2	105.2	1
Madagascar	FBS	598.5	447.2	79.3	188.5	70.6	14.7	17.7	1
Malawi	FBS	505.9	64.5	80.2	86.3	14.9	24.4	21.3	1
Mali	FBS	461.6	45.4	45.7	4.1	53.0	26.6	59.8	1
Mauritius	FBS	545.9	41.8	74.9	31.8	39.1	44.6	267.8	1
Niger	FBS	682.5	93.1	63.4	20.0	56.8	2.8	102.8	1
Niger	INT	635.0	4.2	20.1	0.4	14.1	—	70.9	3
Réunion	FBS	519.1	54.9	71.3	158.5	116.4	63.8	159.0	1
Senegal	FBS	611.9	17.1	42.6	33.8	40.5	69.5	97.5	1
Senegal	WS	420.6	4.8	—	—	25.7	47.4	46.2	4
Senegal	HE	329.4	36.4	116.0	10.5	29.3	147.2	14.6	5
Sierra Leone	FBS	471.0	79.5	114.6	90.6	16.4	50.7	38.9	1
Somalia	FBS	333.2	19.6	15.9	113.3	126.6	5.1	470.9	1
Swaziland	FBS	431.9	63.5	53.1	108.6	106.8	0.0	180.5	1
Zambia	FBS	511.6	92.3	89.2	32.5	40.6	24.5	28.1	1
Zimbabwe	FBS	469.4	26.1	45.2	32.5	40.2	4.9	49.1	1
Average (g/day)		492.7	79.8	60.7	61.9	50.6	32.8	118.9	
Average (kg/year)		179.9	29.1	22.2	22.6	18.5	12.0	43.4	

Table 7b. African-type root- and tuber-based diet (per caput in g/day)

Country or area	Type of data	Cereal	Roots and tubers	Vegetables	Fruit	Meat	Fish	Milk	Reference
Benin	FBS	261.8	767.2	85.7	99.9	21.1	29.3	14.8	1
Burundi	FBS	191.8	1020.3	87.0	224.7	14.9	8.6	48.8	1
Cameroon	FBS	255.2	483.0	110.0	260.7	38.5	30.1	25.9	1
Central African Republic	FBS	112.1	1159.9	47.5	160.1	48.1	15.2	9.9	1
Central African Republic	HE	41.9	328.2	57.5	40.0	76.0	18.3	—	6
Comoros	FBS	340.7	684.3	18.4	257.9	30.0	30.9	30.7	1
Congo	FBS	139.3	1155.4	54.2	313.7	30.0	79.7	26.5	1
Congo	HE	8.9	699.5	131.0	47.3	32.7	34.1	0.0	7
Côte d'Ivoire	FBS	391.7	705.2	102.3	248.2	50.3	55.3	57.6	1
Fiji	FBS	385.2	506.7	84.8	61.5	56.2	119.8	151.4	1
Ghana	FBS	192.7	633.5	82.5	179.9	28.9	63.9	11.7	1
Guinea	FBS	306.8	359.9	191.6	220.5	20.9	17.0	26.2	1
Kiribati	FBS	311.8	527.4	179.6	204.4	41.5	178.3	25.9	1
Liberia	FBS	478.3	483.5	83.2	155.4	30.6	45.6	27.5	1
Mozambique	FBS	222.1	685.3	43.2	71.1	18.7	9.1	25.6	1
Nigeria	FBS	326.8	667.3	98.4	75.5	32.1	44.2	30.5	1
Rwanda	FBS	77.5	988.5	87.6	263.9	16.1	0.7	26.7	1
Samoa	FBS	153.2	597.0	13.3	668.6	130.2	96.6	37.0	1
Sao Tome and Principe	FBS	308.5	419.0	72.5	116.0	20.4	56.4	76.0	1
Solomon Islands	FBS	170.9	897.2	48.9	150.6	30.3	138.2	39.2	1
Togo	FBS	293.6	815.7	66.0	38.0	25.1	29.7	5.7	1
Tonga	FBS	152.3	1393.7	147.8	219.9	135.6	98.7	34.9	1
Uganda	FBS	183.6	326.5	51.0	418.6	32.7	35.5	71.2	1
United Rep. of Tanzania	FBS	215.4	666.5	126.9	236.1	32.1	32.5	70.2	1
Vanuatu	FBS	202.4	422.3	147.1	119.7	112.5	94.8	124.0	1
Zaire	FBS	106.4	1182.4	43.5	193.7	18.7	16.6	7.7	1
Average (g/day)		205.6	658.7	81.3	185.7	41.2	49.4	39.5	
Average (kg/year)		75.0	240.4	29.7	67.8	15.0	18.1	14.4	

Table 7c. Averages for African-type diets (per caput)[a]

	Cereal	Roots and tubers	Vegetables	Fruit	Meat	Fish	Milk
g/day	349.2	369.2	71.0	123.8	45.9	41.1	79.2
kg/year	127.4	134.8	25.9	45.2	16.8	15.0	28.9

[a] North African-type diets are not included (see Table 13).

Table 8. Central American-type diet (per caput in g/day)

Country or area	Type of data	Cereal	Roots and tubers	Vegetables	Fruit	Meat	Fish	Milk	Reference
Antigua	FBS	259.6	35.8	43.1	314.3	83.8	63.1	353.6	1
Bahamas	FBS	244.9	36.6	195.3	181.2	188.4	35.9	254.6	1
Barbados	FBS	323.0	172.6	124.4	157.1	249.0	76.5	235.7	1
Barbados	WS	206.0	145.0	52.0	28.0	135.0	—	114.0	8
Belize	FBS	351.4	226.8	67.5	293.8	97.9	13.0	399.8	1
Bermuda	FBS	228.5	77.6	311.1	264.9	288.8	101.9	355.3	1
Costa Rica	FBS	344.5	31.5	57.0	273.1	90.5	17.1	331.3	1
Cuba	FBS	420.0	204.7	94.1	172.4	104.3	46.7	422.3	1
Dominica	FBS	232.0	364.3	138.3	519.2	73.3	58.5	168.5	1
Dominican Republic	FBS	279.2	76.5	80.1	534.0	64.7	21.0	217.6	1
Grenada	FBS	247.8	83.9	56.9	337.1	88.0	90.8	228.5	1
Guadeloupe	FBS	368.4	124.5	183.4	213.7	136.9	133.6	246.7	1
Guatemala	FBS	373.4	14.6	55.8	122.2	54.9	2.2	100.9	1
Haiti	FBS	256.9	243.0	119.9	332.9	34.2	8.5	34.5	1
Honduras	FBS	357.6	15.4	47.0	280.8	36.7	3.7	118.5	1
Jamaica	FBS	320.9	233.1	88.3	246.2	104.8	49.6	78.2	1
Martinique	FBS	336.8	151.7	282.7	284.5	145.0	127.5	181.0	1
Mexico	FBS	475.1	37.5	86.6	251.0	69.6	28.8	274.5	1
Netherlands Antilles	FBS	309.3	86.6	111.4	243.3	199.0	53.2	420.7	1
Panama	FBS	344.6	92.2	58.6	285.0	108.9	19.8	168.6	1
Saint Kitts and Nevis	FBS	206.1	149.1	76.6	104.1	107.6	83.4	138.6	1
Saint Lucia	FBS	253.2	230.4	33.4	715.3	138.9	83.6	160.7	1
Saint Vincent	FBS	250.3	44.5	16.4	226.6	80.0	33.9	166.1	1
Trinidad and Tobago	FBS	437.0	108.1	107.8	197.0	114.3	33.8	345.2	1
Trinidad and Tobago	INV	325.0	161.0	171.0	173.0	103.0	40.0	115.0	8
Average (g/day)		310.1	125.9	106.4	270.0	115.9	51.1	225.2	
Average (kg/year)		113.2	46.0	38.8	98.6	42.3	18.7	82.2	

Table 9. Chinese-type diet (per caput in g/day)

Country or area	Type of data	Cereal	Roots and tubers	Vegetables	Fruit	Meat	Fish	Milk	Reference
China	FBS	627.1	300.9	180.7	30.0	62.7	15.6	13.4	1
China	DS	451.0	—	286.0	0.0	45.0	38.0	2.0	9
Viet Nam	FBS	387.0	273.0	—	—	43.5	4.6	5.4	10
Viet Nam	DS	416.0	131.0	—	—	13.0	39.0	0.0	10
Average (g/day)		470.3	235.0	233.4	15.0	41.1	24.3	5.2	
Average (kg/year)		171.7	85.8	85.2	5.5	15.0	8.9	1.9	

Table 10. Eastern Mediterranean-type diet (per caput in g/day)

Country or area	Type of data	Cereal	Roots and tubers	Vegetables	Fruit	Meat	Fish	Milk	Reference
Bangladesh	FBS	631.7	44.8	26.9	39.6	10.8	20.3	36.7	1
Democratic Yemen	FBS	469.5	6.3	80.3	206.3	34.5	50.4	204.8	1
Egypt	FBS	694.3	64.3	328.6	208.3	42.0	13.6	57.4	1
Kuwait	FBS	469.2	49.2	329.6	374.3	217.0	28.2	457.4	1
Saudi Arabia	FBS	454.7	28.0	220.7	520.7	133.1	25.1	319.0	1
Sudan	FBS	390.2	40.9	93.1	112.0	69.8	4.1	179.4	1
Syrian Arab Republic	FBS	525.7	68.3	563.0	474.6	61.2	4.7	184.8	1
Turkey	FBS	576.6	140.0	343.8	462.0	61.1	17.8	175.0	1
Turkey	RC	544.8	51.0	193.1	79.4	36.8	8.6	107.8	11
United Arab Emirates	FBS	366.0	40.3	467.3	449.4	188.6	67.7	378.7	1
Yemen	FBS	550.1	47.5	112.7	130.9	60.1	11.7	138.9	1
Average (g/day)		515.7	52.8	250.8	278.0	83.2	22.9	203.6	
Average (kg/year)		188.2	19.3	91.6	101.5	30.4	8.4	74.3	

Table 11. European-type diet (per caput in g/day) *(continues on page 42)*

Country or area	Type of data	Cereal	Roots and tubers	Vegetables	Fruit	Meat	Fish	Milk	Reference
Australia	FBS	311.2	136.9	182.7	247.7	283.0	39.0	331.3	1
Australia	RC	232.0	—	268.0	178.0	192.0	19.0	427.0	19
Austria	FBS	255.7	229.1	218.6	315.0	265.3	24.3	417.7	1
Belgium and Luxembourg	FBS	265.3	273.9	231.3	229.7	267.9	51.5	361.0	1
Bulgaria	FBS	616.9	77.7	270.3	312.3	185.2	15.2	243.2	1
Canada	FBS	253.7	218.3	260.6	326.0	262.0	60.7	468.3	1
Czechoslovakia	FBS	399.1	217.8	187.2	141.3	260.9	22.3	426.2	1
Denmark	FBS	239.3	207.4	137.4	172.3	220.0	133.4	432.2	1
Denmark	HE?	264.0	213.0	—	—	149.0	27.0	681.0	12
Faroe Islands	FBS	367.1	247.0	73.5	157.7	220.9	272.0	203.0	1
Finland	FBS	266.7	238.4	87.1	219.1	168.9	78.4	711.2	1
France	FBS	290.2	218.6	305.6	187.0	273.7	66.0	346.6	1
France	HE?	233.0	—	397.0	196.0	187.0	26.0	294.0	13
German Democratic Republic	FBS	362.8	389.1	206.0	195.5	293.1	38.1	389.3	1
Germany, Federal Republic of	FBS	254.3	221.1	187.5	104.8	267.7	27.2	328.7	1
Greece	FBS	414.1	182.2	495.7	453.7	185.7	44.7	262.1	1
Hungary	FBS	423.8	162.3	226.8	217.7	284.4	9.9	315.3	1
Iceland	FBS	226.8	171.4	60.5	160.0	250.1	291.2	695.9	1
Ireland	FBS	366.5	313.7	255.6	180.2	228.5	43.2	637.2	1
Ireland	HE	318.4	59.9	—	—	—	—	456.0	14
Israel	FBS	394.1	113.1	296.4	391.3	190.8	46.8	315.3	1
Israel	HE	215.0	67.0	232.0	297.0	99.0	17.0	400.0	16
Israel	FBS	306.5	107.3	318.9	400.4	178.6	23.3	330.0	17
Italy	FBS	503.1	112.8	457.0	353.8	200.3	34.2	299.7	1
Malta	FBS	391.3	62.8	261.8	176.4	154.7	57.5	416.8	1
Netherlands	HE?	329.0	276.0	210.0	256.0	—	9.0	602.0	15
New Zealand	FBS	291.8	157.9	232.3	212.3	320.8	27.2	856.9	1
Norway	FBS	296.3	233.5	120.5	220.3	146.5	141.2	692.6	1
Norway	HE	203.2	151.0	87.2	131.8	79.7	54.1	460.1	14
Poland	FBS	495.2	326.3	309.5	102.1	193.3	46.0	437.2	1
Poland	HE	318.4	291.5	185.5	105.2	148.8	15.6	389.9	14
Portugal	FBS	442.5	265.8	337.7	191.0	139.4	77.0	146.8	1
Romania	FBS	523.9	193.3	382.8	154.6	189.2	16.7	403.7	1

Table 11. European-type diet (per caput in g/day) *(continued)*

Country or area	Type of data	Cereal	Roots and tubers	Vegetables	Fruit	Meat	Fish	Milk	Reference
Spain	FBS	324.6	306.9	405.0	347.8	187.2	86.8	328.7	1
Spain	HE	245.0	230.8	213.5	248.0	170.5	71.7	489.0	14
Sweden	FBS	230.5	202.2	129.8	218.8	179.7	83.3	502.0	1
Switzerland	FBS	271.9	133.4	191.9	312.7	239.6	29.4	460.9	1
United Kingdom	FBS	258.9	281.2	197.5	151.9	205.2	45.1	455.0	1
United Kingdom	HE	226.3	169.9	181.5	113.1	159.8	20.0	367.0	14
USA	FBS	250.0	146.7	271.6	309.3	312.2	44.2	462.3	1
USSR	FBS	504.1	300.5	258.9	106.8	169.7	69.7	382.1	1
Yugoslavia	FBS	599.4	158.7	217.2	194.8	176.7	8.5	298.3	1
Yugoslavia	HE	285.0	98.0	184.0	148.0	175.0	13.0	330.0	18
Average (g/day)		331.8	199.1	237.4	222.9	206.4	55.4	424.5	
Average (kg/year)		121.1	72.7	86.7	81.4	75.3	20.2	154.9	

Table 12. Far Eastern-type diet (per caput in g/day)

Country or area	Type of data	Cereal	Roots and tubers	Vegetables	Fruit	Meat	Fish	Milk	Reference
Brunei Darussalam	FBS	509.4	73.5	87.5	181.4	95.3	94.5	161.2	1
Burma	FBS	799.2	10.1	130.1	73.8	17.4	39.2	14.4	1
Hong Kong	FBS	387.4	35.3	253.4	180.8	203.4	135.5	95.2	1
India	FBS	502.2	53.6	160.4	63.4	3.7	8.4	104.7	1
India	WS	498.0	47.0	76.0	21.0	5.0	10.0	78.0	20
Indonesia	FBS	597.0	197.2	34.8	50.9	9.4	31.9	10.3	1
Indonesia	HE	624.0	284.1	147.8	100.4	32.3	43.4	20.1	21
Japan	FBS	502.8	72.0	299.1	178.2	82.4	231.8	135.5	1
Macao	FBS	437.8	17.8	202.1	141.5	170.3	130.9	45.7	1
Malaysia	FBS	536.8	61.5	91.4	130.2	51.5	123.6	55.4	1
Nepal	FBS	575.1	47.2	30.2	23.0	14.9	0.9	115.1	1
Pakistan	FBS	462.0	14.9	58.1	66.5	26.3	4.7	192.0	1
Pakistan	FBS	412.3	15.1	69.5	78.1	24.9	3.5	228.7	17
Pakistan	HE	409.0	24.0	63.1	70.0	20.4	2.0	170.7	14
Philippines	FBS	569.2	176.4	94.5	266.2	44.0	86.1	48.1	1
Philippines	WS	367.0	37.0	145.0	142.0	54.0	8.0	33.0	22
Republic of Korea	FBS	865.2	78.2	520.2	95.8	43.4	138.9	32.4	1
Singapore	FBS	559.3	84.9	190.0	219.8	174.5	86.6	113.1	1
Sri Lanka	FBS	511.0	94.1	47.4	215.8	6.6	38.6	72.0	1
Thailand	FBS	631.3	53.4	123.2	216.4	40.0	52.6	21.6	1
Average (g/day)		566.1	77.8	148.6	132.4	58.9	66.9	91.0	
Average (kg/year)		206.6	28.4	54.3	48.3	21.5	24.4	33.6	

Table 13. North African-type diet (per caput in g/day)

Country or area	Type of data	Cereal	Roots and tubers	Vegetables	Fruit	Meat	Fish	Milk	Reference
Algeria	FBS	527.2	80.9	104.9	142.9	30.7	5.7	221.0	1
Libyan Arab Jamahiriya	FBS	569.3	80.4	265.7	402.3	145.2	22.0	280.2	1
Mauritania	FBS	383.0	14.5	10.1	30.7	82.0	44.7	421.1	1
Morocco	FBS	589.0	49.0	107.0	111.7	41.2	15.2	85.9	1
Morocco	HE	592.9	—	243.0	127.4	49.0	9.9	81.1	23
Tunisia	FBS	531.7	49.2	312.1	235.8	46.6	21.9	179.6	1
Average (g/day)		443.6	54.8	173.8	175.1	65.8	19.9	211.5	
Average (kg/year)		161.9	20.0	63.4	63.9	24.0	7.3	77.2	

Table 14. South American-type diet (per caput in g/day)

Country or area	Type of data	Cereal	Roots and tubers	Vegetables	Fruit	Meat	Fish	Milk	Reference
Argentina	FBS	380.2	212.0	181.6	288.9	346.5	16.0	269.6	1
Bolivia	FBS	298.5	335.2	122.6	195.6	90.4	7.9	83.7	1
Brazil	FBS	371.3	264.0	75.9	205.6	96.2	19.4	183.2	1
Brazil	HE	224.0	99.0	47.0	53.0	—	—	138.0	24
Chile	FBS	466.8	127.4	188.6	183.3	92.6	78.1	250.1	1
Colombia	FBS	295.5	297.7	118.6	272.3	96.5	11.3	168.1	1
Ecuador	FBS	253.9	131.5	75.2	447.6	78.4	47.0	220.9	1
French Guiana	FBS	361.7	148.1	123.0	266.0	215.3	80.0	163.7	1
Guyana	FBS	514.6	37.0	24.9	114.9	61.5	65.6	171.7	1
Guyana	WS	322.0	95.0	91.0	39.0	93.0	—	70.0	8
Paraguay	FBS	287.1	515.5	80.6	586.4	197.4	2.3	140.5	1
Peru	FBS	339.6	275.7	90.1	206.9	64.2	81.5	130.1	1
Suriname	FBS	522.6	50.4	56.1	81.6	105.4	61.7	109.9	1
Uruguay	FBS	368.4	140.5	123.0	172.0	255.0	20.3	441.8	1
Venezuela	FBS	353.2	80.9	58.4	325.3	133.3	30.2	366.4	1
Venezuela	HE	317.4	155.2	41.6	208.2	62.4	29.2	190.1	25
Average (g/day)		354.8	185.3	93.6	227.9	132.5	39.3	193.6	
Average (kg/year)		129.5	67.6	34.2	83.2	48.4	14.4	70.7	

References

1. *Food balance sheets, 1979–1981 averages.* Rome, FAO, 1984.
2. SHAH, M.M. & FROBERG, H. *Food consumption patterns — rural and urban Kenya.* Laxenburg, Austria, International Institute for Applied Systems Analysis, 1980 (working paper WP-80-14).
3. *Aperçu de la consommation alimentaire et de l'état nutritionelle des populations rurales de l'arrondissement de Keïta-Niger (1984). [General survey of food consumption and nutritional state of the rural populations of the district of Keïta-Niger (1984).]* Rome, FAO, 1985 (document GCP/NER/028/ITA).
4. BENEFICE, E. ET AL. Etudes de nutrition dans la Moyenne Vallée du Sénégal. 1. Evolution de la consommation alimentaire depuis 1958 et structure actuelle de la ration. [Nutritional studies in the mid-Valley of Senegal river. 1. Food consumption evolution since 1958 and current structure of food intake.] *Bulletin de la Société de Pathologie exotique et de ses filiales,* **78**: 110–118 (1985).
5. CANNONE, P.J. & CHEVASSUS-AGNÈS, S.P. *Rapport final de l'enquête de la consommation alimentaire et nutritionelle effectué à Dakar (juin–juillet 1977). [Final report of the food consumption and nutrition inquiry carried out in Dakar, June–July 1977.]* Dakar, Organisme de Recherche pour L'Alimentation et la Nutrition en Afrique, 1978.
6. BUI QUANG MINH. *Enquête budget et consommation des ménages en Empire Centrafricain. Annex II: résultats statistiques. [Enquiry on budget and household consumption in the Central African Empire. Annex II: statistical results.]* Bangui, United Nations Development Programme, June 1977.
7. CRESTA, M. ET AL. L'économie paysanne et l'alimentation dans les villages de Oka-Bamboo (Ewo) et de Inkala-Matiba (Kindaba). [Farming economy and diet in the villages of Oka-Bamboo (Ewo) and Inkala-Matiba (Kindaba).] In: *Recherches biologiques, nutritionelles et sanitaires sur des populations de la République Populaire du Congo et problèmes liés au développement rural. [Biological, nutritional and sanitary research on the populations of the People's Republic of the Congo and problems linked to rural development.]* Rome, Instituto Italiano di Antropologia, 1984/1985.
8. SEN GUPTA, P.N. Comparative studies on food consumption and nutrition in three Caribbean countries: Barbados, Trinidad and Tobago and Guyana. *Ecology of food and nutrition,* **11**: 177–189 (1981).
9. SHAO YU-FENG ET AL. A dietary survey of Qi–bao middle school students. *American journal of public health,* **72**(suppl.): 43 (1982).
10. AUTRET M. *La situation alimentaire et nutritionelle au Viet-Nam 1983. [The diet and nutrition situation in Viet Nam 1983.]* Viet Nam, United Nations Children's Fund (unpublished document).
11. GENÇAGA, H. *Food consumption and nutrition in Turkey.* Ankara, Turkish Development Research Foundation, 1985 (Publication No. 6).
12. ANDERSEN, A. *Bly, cadmium, kobber og zink in den danske kost. [Lead, cadmium, copper and zinc in the Danish diet.]* Denmark, Statens

Levnedsmiddelinstitut (no year).

13. DESCHAMPS, P. & HASCOET, M. National survey of the food quality in France. In: Miyamoto, J. et al., ed. *IUPAC pesticide chemistry.* Oxford, Pergamon, 1983.

14. *Review of food consumption surveys — 1985.* Rome, FAO, 1986 (FAO Food and Nutrition Paper No. 35).

15. Unpublished information supplied by the State Department of Public Health, The Netherlands, 1987.

16. BAVLY, S. ET AL. *Levels of nutrition in Israel 1975/76.* Jerusalem, Ministry of Education and Culture/Hebrew University, Hadassah Medical School, 1980.

17. *Food balance sheets, 1975–1977 average, and per caput food supplies, 1961–1965, average 1966–1977.* Rome, FAO, 1980.

18. *Statistical yearbook of the city of Zagreb 1982.*

19. COMMONWEALTH DEPARTMENT OF HEALTH. *National dietary survey of adults 1983. No. 1 — foods consumed.* Canberra, Australian Government Publishing Service, 1986.

20. NATIONAL NUTRITION MONITORING BUREAU. *Report for the year 1982.* Hyderabad, National Institute of Nutrition, 1984.

21. WORLD BANK. *Patterns of food consumption and nutrition in Indonesia. An analysis of the National Socio-economic Survey (SUSENAS) 1978.* Washington, DC, World Bank (Staff Working Paper No. 670).

22. FOOD AND NUTRITION RESEARCH INSTITUTE. *First nationwide nutrition survey, Philippines 1978. First revision April, 1980.* Manila, Food and Nutrition Research Institute, 1980 (Publication No. GP-11).

23. ESSATARA, M. *Situation alimentaire et nutritionelle au Maroc. 1ère conférence nationale sur l'alimentation et la nutrition, 13–15 février 1978. [Diet and nutrition situation in Morocco. First national conference on diet and nutrition, 13–15 February 1978.]* Rabat, Commission interministérielle pour l'Alimentation et la Nutrition, 1978.

24. KRUSE, B. *Padroes e habitos alimentares da populaçao brasileira. [Dietary patterns and habits of the Brazilian population.]* Rio de Janeiro, Instituto Nacional de Alimentaçao e Nutriçao, 1981.

25. BERMUDES CHAURIO, L. ET AL. *Atlas para nutrición. [Atlas for nutrition.]* Caracas, Instituto Nacional de Nutrición, 1974.

Annex 3

RADIATION DOSE PER UNIT INTAKE

ICRP (1979, 1980) has recommended dosimetric models and metabolic data based on "reference man" for estimating radiation doses from intakes of radionuclides. These models and data have been developed for adults exposed to radiation at work, and ICRP does not recommend their use for calculating the committed dose equivalent for individual members of the public from the intake of radionuclides in the environment. Several factors will influence the average dose to a mixed population.

Body size and biokinetics

Even if there were no differences between children and adults in the uptake and retention of a radionuclide, the dose equivalent in a particular tissue per unit intake of the radionuclide would be greater in the former, because of the smaller mass of their organs and tissues. For short-lived radionuclides emitting low-penetrating radiation (β-particles, α-particles, photons with energies below 10 keV), the dose equivalent per unit intake will be greater in children than in adults according to the inverse ratio of organ or tissue masses. For long-lived radionuclides emitting low-penetrating radiation, which are retained longer in the body, this ratio will be only about 2, because, as the mass of an organ or tissue increases with age, the activity concentration of the retained radionuclide decreases. For radionuclides emitting penetrating radiation (photons with energies above 10 keV) the modifying factor for body size is smaller, because the dose per unit intake factor in a particular organ is less dependent on the mass of the organ. Even if only differences in body size are allowed for, the committed dose equivalents per unit intake calculated for young members of the public will therefore be greater than those for adults, by factors ranging from less than 2 to 10, depending on the type of radiation emitted by the radionuclide and its effective half-life in the body organs or tissues.

The biokinetics of radionuclides may also differ substantially between children and adults. This may result in a larger fraction of the radionuclide being deposited in the organs or tissues of children or, frequently, in the more rapid elimination of a radionuclide by children. For example, since caesium-137 has a more rapid turnover in children, the dose equivalent in body tissues from a unit intake of caesium-137 for a one-year-old is only about three-quarters that for an adult.

The biological half-life of iodine in the thyroid increases with age, but the deposition of iodine in the thyroid is slightly higher in the first months of life than in the young child, adolescent, and adult. However, because of the comparatively short half-life of iodine-131, age-related differences in biological turnover are of little consequence to the effects of this radionuclide, so that the greater thyroid mass in adults (10 times that of infants) is the major factor influencing the ratio of the dose equivalent per unit intake for the young child to that for the adult.

The committed dose equivalent per unit intake of long-lived strontium-90 for the six-month-old child is about five times the adult value. However, for strontium-89, which has a much shorter half-life, the corresponding ratio lies in the range 20–40, depending on the "bone model" used for calculations.

The differences between children and adults in the biokinetics of radionuclides (and in the mass of organs and tissues) are reflected in organ dose factors. For an infant these range from 0.7 to 40 times the values for the adult.

Gastrointestinal absorption

Animal experiments have shown that the absorption of radionuclides from the gastrointestinal tract is higher in the newborn than in adults, but falls to adult values by about the time of weaning. Although the extent of absorption in the very young is dependent on the radionuclide considered, in general, the smaller the rate of gastrointestinal absorption in adults, the greater is the ratio of the rate in the newborn to that in the adult. If, for a six-month-old infant, the age-dependency of absorption of plutonium-239 from the gastrointestinal tract is taken into account, and the body-mass-dependent factor of 2 for long-retained radionuclides is applied, the committed dose equivalent per unit intake of dietary plutonium-239 is about 20 times that for the adult.

The dose per unit intake factors for calculating the Annual Limits on Intake given by ICRP (1979, 1980) are usually appropriate for the chemical forms of a radionuclide most likely to be encountered in the workplace. Chemical forms of the same radionuclide found in the environment, or in food, may differ markedly from these, and may therefore exhibit a different biokinetic behaviour. Other factors that may influence the absorption of radionuclides from the gastrointestinal tract are nutritional status, valence, and the presence of other elements in foodstuffs that could compete with transport mechanisms. Consequently, the dose equivalent per unit intake is liable to vary, especially if the absorption of the radionuclide in the upper gastrointestinal tract is enhanced, which will decrease the absorbed dose per unit intake to the large intestine.

Exposure of the fetus

ICRP has so far made no recommendations on methods to be used for calculating radiation doses to the fetus after intakes of radionuclides by the mother. A number of dosimetric models have been published for specific radionuclides. However, human data are available only for a few radionuclides such as caesium-137, strontium-90, and iodine-131.

In the case of intakes of iodine-131 by pregnant women, the value of the dose equivalent per unit intake for the fetal thyroid is of special concern. Before week 12 after conception, no iodine accumulation occurs in the fetal thyroid; therefore the dose per unit intake factor for iodine-131 before week 12 is comparable to that for maternal soft-tissue, which is orders of magnitude below that for the maternal thyroid. After the first trimester, the fetal thyroid gradually develops and increases in function, so that the dose per unit intake factor for the fetal thyroid is approximately half that for the maternal thyroid in the second trimester, and approximately equal to it towards the end of pregnancy.

Exposure of children

For the 10-year-old age group, the dose factors for caesium-134 and caesium-137 are lower than the adult values. The factors for strontium-90 and plutonium-239 are marginally higher than for the adult, but only to a degree that should be compensated for by the lower food consumption rates associated with the younger age group; that is, the smaller intake of radioactive contamination will balance out the higher dose per unit intake factors. For these four nuclides, therefore, derived intervention levels designed to protect adults will also protect the 10-year-old age group.

For iodine-131, however, the adult dose per unit intake factors are less than half of those for 10-year-olds, which in turn are lower than those for infants. The application of the derived intervention level for infants would give more than adequate protection not only to adults but also to the 10-year-old age group.

Conclusions

Except for a limited number of elements, such as strontium, iodine, and caesium, attempts to extend the "reference man" dosimetric and metabolic criteria to members of the general public are hampered by lack of information regarding the biokinetic behaviour of ingested or inhaled radionuclides. With some exceptions, a conservative approach, in the absence of relevant information on age-dependent biokinetics, is to use the metabolic data for the adult for the derivation of dose per unit intake factors.

The dose per unit intake factors for different radionuclides listed in Table 4 (page 25) are based on such considerations and on a review of the age-dependent dose per unit intake factors published by CEC, IAEA, the Institute for Radiation Hygiene, Federal Health Office, Federal Republic of Germany, and the National Radiological Protection Board, England. The values in Table 4 should be used when critical-group calculations are undertaken. For calculating guideline values for derived intervention levels for the general population, however, the two rounded-off dose per unit intake factors of 10^{-6} and 10^{-8} Sv/Bq have been used here (Table 5, page 26).

References

ICRP. Limits for intakes of radionuclides by workers. *Annals of the ICRP,* **2**(3/4) (1979) (ICRP Publication 30, Part 1).

ICRP. Statement and recommendations of the 1980 Brighton meeting of the ICRP. Limits for intakes of radionuclides by workers. *Annals of the ICRP,* **4**(3/4) (1980) (ICRP Publication 30, Part 2).

SAMPLE CALCULATIONS OF DERIVED INTERVENTION LEVELS

The derived intervention levels presented in Table 5 (page 26) have been calculated on the basis of a single radionuclide in a single foodstuff leading to the intervention level of dose. The recommendation is made that, if more than one food category is affected, or if there are several radionuclides present, modified derived intervention levels should be calculated according to the additivity formula

$$\sum_{i} \sum_{f} \frac{C(i, f)}{DIL(i, f)} \leqq 1 \qquad (1)$$

where $C(i, f)$ is the activity concentration of nuclide i in foodstuff f and $DIL(i, f)$ is the derived intervention level calculated on the assumption that only nuclide i is present and only in foodstuff f. The effect of using the additivity formula is to control the radiation dose to individuals so that it does not exceed the intervention level of dose (5 mSv effective dose equivalent).

In any given situation there will be many ways of meeting the dose criterion. In this Annex a number of examples are given in which derived intervention levels are calculated for various combinations of radionuclide concentrations in several food categories. Before these situation-specific derived intervention levels (DIL^* values) can be calculated, the inequality in equation (1) needs to be transformed. It can be shown that, for a given set of contamination assumptions, the resulting DIL^* for nuclide i in foodstuff f will be

$$DIL^*(i, f) = \frac{g(i, f)}{\sum_{i} \sum_{f} \frac{g(i, f)}{DIL(i, f)}} \qquad (2)$$

where $g(i, f)$ is a function that represents the specific pattern of contamination. As $g(i, f)$ appears in both the numerator and the denominator of equation (2), it can be expressed simply in relative terms, i.e., as the ratio of activity concentrations found in different foodstuffs, or the ratio of activities of different radionuclides. This is illustrated in the examples below.

It should be emphasized that, although the results are presented with arithmetic accuracy, they should be rounded off before application, as in Example 1.

Example 1. Base case: all foodstuffs contaminated with caesium-137

In this case the total diet is taken to be contaminated with a single radionuclide, caesium-137, which has a dose per unit intake factor of 10^{-8} Sv/Bq. Food intake is taken as 550 kg per year, so that the *DIL** leading to 5 mSv is given by

$$5 \times 10^{-3}\ (\text{Sv/year}) = 550\ (\text{kg/year}) \times DIL^*\ (\text{Bq/kg}) \times 10^{-8}\ (\text{Sv/Bq})$$

$$\therefore \quad DIL^* = \frac{5 \times 10^{-3}}{550 \times 10^{-8}}\ \text{Bq/kg}$$

$$DIL^* = 909\ \text{Bq/kg}.$$

Thus if all foodstuffs are uniformily contaminated with caesium-137, the specific derived intervention level can be taken as 1000 Bq/kg, which would also be applicable to any radionuclide or mixture of radionuclides whose dose per unit intake factor were 10^{-8} Sv/Bq.

Example 2. Two food categories contaminated with caesium-137

In this example it is assumed that caesium-137 has been found in two food categories, meat and milk. The general derived intervention levels for caesium-137 are (from Table 5, page 26):

meat	10 000	Bq/kg
milk	4500	Bq/kg.

For the first calculation it is assumed that the activity concentration found is the *same* in both meat and milk. The ratio of the $g(i,f)$ values for equation (2) can be taken as 1 for both milk and meat.

Thus for both milk and meat

$$DIL^* = \frac{1}{\frac{1}{10\,000} + \frac{1}{4500}} = 3103 \quad \text{Bq/kg.}$$

For the second calculation, it is assumed that the activity concentration in meat is four times that in milk. Thus the relative g(*i,f*) for meat is 4 and that for milk is 1, which gives

$$DIL^*(meat) = \frac{4}{\frac{4}{10\,000} + \frac{1}{4500}} = 6429 \quad \text{Bq/kg}$$

$$DIL^*(milk) = \frac{1}{\frac{4}{10\,000} + \frac{1}{4500}} = 1607 \quad \text{Bq/kg.}$$

The *DIL** values from both calculations can be compared with the derived intervention levels quoted for milk and meat in Table 5 (page 26).

Example 3. Three food categories contaminated with one radionuclide

In this example it is assumed that meat, milk, and cereals are contaminated with caesium-137. The reference derived intervention levels from Table 5 (page 26) are:

meat	10 000	Bq/kg
milk	4500	Bq/kg.
cereals	3500	Bq/kg.

For the first application it is assumed that milk and cereals are contaminated at the same activity concentration of caesium-137 and that meat has twice that concentration. Thus the relative g(*i,f*) values are:

meat	2
milk	1
cereals	1

and the *DIL** values are given by

$$DIL^*(meat) = \frac{2}{\frac{2}{10\,000} + \frac{1}{4500} + \frac{1}{3500}} = 2825 \quad \text{Bq/kg}$$

$$DIL^*(milk) = \frac{1}{\frac{2}{10\,000} + \frac{1}{4500} + \frac{1}{3500}} = 1412 \text{ Bq/kg}.$$

The *DIL*(cereal)* is 1412 Bq/kg because the same activity concentration was assumed as for milk.

If the relative activity concentrations are now given as:

meat	7
milk	2
cereal	1

the *DIL** values are

$$DIL^*(meat) = \frac{7}{\frac{7}{10\,000} + \frac{2}{4500} + \frac{1}{3500}} = 4895 \text{ Bq/kg}$$

$$DIL^*(milk) = \frac{2}{\frac{7}{10\,000} + \frac{2}{4500} + \frac{1}{3500}} = 1398 \text{ Bq/kg}$$

$$DIL^*(cereal) = \frac{1}{\frac{7}{10\,000} + \frac{2}{4500} + \frac{1}{3500}} = 699 \text{ Bq/kg}.$$

Restricting the activity concentrations of caesium-137 in these three foodstuffs to below these *DIL** levels would ensure that no individual received a dose of 5 mSv from intakes in the first year after the accident.

Example 4. One foodstuff contaminated with two radionuclides

In this example it is assumed that milk is the only foodstuff affected, but that both iodine-131 and caesium-137 are present. Calculations have been done for three ratios of activity concentration of iodine-131 to caesium-137:

(*a*) 10 : 1
(*b*) 3 : 1
and (*c*) 1 : 10.

This covers the range of possibilities from there being 10 times as much iodine-131 as caesium-137, to there being 10 times as much caesium-137 as iodine-131. The reference derived intervention levels are taken as 1600 Bq/l for iodine-131 (the value for infants) and 4500 Bq/l for caesium-137.

$$(a)\ DIL^*(I\text{-}131,\ milk) = \frac{10}{\frac{10}{1600} + \frac{1}{4500}} = 1545\ \text{Bq/l}$$

$$DIL^*(Cs\text{-}137,\ milk) = \frac{1}{\frac{10}{1600} + \frac{1}{4500}} = 155\ \text{Bq/l}$$

$$(b)\ DIL^*(I\text{-}131,\ milk) = \frac{3}{\frac{3}{1600} + \frac{1}{4500}} = 1430\ \text{Bq/l}$$

$$DIL^*(Cs\text{-}137,\ milk) = \frac{1}{\frac{3}{1600} + \frac{1}{4500}} = 477\ \text{Bq/l}$$

$$(c)\ DIL^*(I\text{-}131,\ milk) = \frac{1}{\frac{1}{1600} + \frac{10}{4500}} = 351\ \text{Bq/l}$$

$$DIL^*(Cs\text{-}137,\ milk) = \frac{10}{\frac{1}{1600} + \frac{10}{4500}} = 3512\ \text{Bq/l}$$

From these three examples it can be seen that for a given radionuclide the *DIL** value moves closer to the reference value as the nuclide becomes dominant in terms of activity concentration.

Example 5. Two foodstuffs contaminated with two radionuclides

Here the assumption is made that cereals and meat are contaminated by plutonium-239 and caesium-137 in different relative activity concentrations. Suppose that the relative activity concentrations are:

	Pu-239	Cs-137
meat	1	10 000
cereals	10	1000.

The reference derived intervention levels from Table 5 (page 26) are for meat: plutonium-239 100 Bq/kg, caesium-137 10 000 Bq/kg; and for cereals: plutonium-239 35 Bq/kg, caesium-137 3500 Bq/kg. Consequently

$$DIL^*(Pu\text{-}239,\ meat) = \frac{1}{\frac{1}{100}+\frac{10}{35}+\frac{1000}{3500}+\frac{10\,000}{10\,000}} = 0.63 \text{ Bq/kg}$$

$$DIL^*(Pu\text{-}239,\ cereals) = \frac{10}{\frac{1}{100}+\frac{10}{35}+\frac{1000}{3500}+\frac{10\,000}{10\,000}} = 6.3 \text{ Bq/kg}$$

$$DIL^*(Cs\text{-}137,\ meat) = \frac{10\ 000}{\frac{1}{100}+\frac{10}{35}+\frac{1000}{3500}+\frac{10\,000}{10\,000}} = 6323 \text{ Bq/kg}$$

$$DIL^*(Cs\text{-}137,\ cereals) = \frac{1000}{\frac{1}{100}+\frac{10}{35}+\frac{1000}{3500}+\frac{10\,000}{10\,000}} = 632 \text{ Bq/kg.}$$

Finally, for relative activity concentrations of:

	Pu-239	Cs-137
meat	10	1000
cereals	1	100

the denominator in the equation is

$$\frac{1}{35}+\frac{10}{100}+\frac{100}{3500}+\frac{1000}{10\,000} = 0.257$$

and the resulting *DIL** values are (in Bq/kg):

	Pu-239	Cs-137
meat	39	3889
cereals	3.9	389.

Annex 5

WHO MEETINGS ON DERIVED INTERVENTION LEVELS

Working Group on Guideline Values for Derived Intervention Levels

Geneva, 6–9 April 1987

Participants

Dr R.M. Bahudarov, Institute of Biophysics, Ministry of Health, Moscow, USSR

Dr D.J. Beninson, National Commission of Atomic Energy, Buenos Aires, Argentina (*Chairman*)

Dr R. Clarke, National Radiological Protection Board, Didcot, Oxon, England

Mr B. Emmerson, International Atomic Energy Agency, Vienna, Austria

Mr P. François, Food and Agriculture Organization of the United Nations, Rome, Italy

Mr G. Fraser, Commission of the European Communities, Luxembourg

Dr L.D. Hamilton, Brookhaven National Laboratory, NY, USA

Dr O. Ilari, Nuclear Energy Agency, Organisation for Economic Co-operation and Development, Paris, France

Dr A. Kaul, Institute for Radiation Hygiene, Federal Health Office, Bad Neuherberg, Federal Republic of Germany

Dr W. Keller, Geneva, Switzerland (*Consultant*)

Mr J.E. Logsdon, Environmental Protection Agency, Washington, DC, USA

Mr J.I. Waddington, Director, Environmental Health, WHO Regional Office for Europe, Copenhagen, Denmark

Dr P.J. Waight, Prevention of Environmental Pollution, WHO, Geneva, Switzerland (*Secretary*)

Task Group on Guideline Values for Derived Intervention Levels

Geneva, 21–25 September 1987

Participants*

Dr I. Al-Rawi, Radiation and Nuclear Medicine Hospital, Baghdad, Iraq

Dr D. Beninson, National Commission of Atomic Energy, Buenos Aires, Argentina (*Chairman*)

Dr R. Gill, Center for Food Safety and Applied Nutrition, Washington, DC, USA

Dr H. Jammet, Atomic Energy Commission, Fontenay-aux-Roses, France

Dr E. Létourneau, Radiation Protection Bureau, Ottawa, Canada (*Vice-Chairman*)

Dr B. Michaud, Federal Office of Public Health, Bern, Switzerland

Mr K. Narayanan, Bhabba Atomic Research Centre, Bombay, India

Professor R. Ndoye, Faculty of Medicine, Dakar, Senegal

Dr E. Rubery, Department of Health and Social Security, London, England (*Rapporteur*)

Dr G. Schmidt, Food and Drug Administration, Rockville, MD, USA

Mr T. Swindon, Australian Radiation Laboratory, Yallambie, Australia

Representatives of other organizations†

Mr L. Chamney, Nuclear Energy Agency, Organisation for Economic Co-operation and Development, Paris, France

Mr G. Fraser, Commission of the European Communities, Luxembourg

Mr J. Lupien, Food and Agriculture Organization of the United Nations, Rome, Italy

Ms A. Salo, International Atomic Energy Agency, Vienna, Austria

---

* Unable to attend: Dr. V. Lyscov, Moscow Engineering Institute, Moscow, USSR.

† The International Commission on Radiological Protection was represented by Dr D. Beninson.

Secretariat

Dr R.H. Clarke, National Radiological Protection Board, Didcot, Oxon, England (*Temporary Adviser*)

Dr L. Hamilton, Brookhaven National Laboratory, NY, USA (*Temporary Adviser*)

Mr J. Waddington, Director, Environmental Health, WHO Regional Office for Europe, Copenhagen, Denmark

Dr P. Waight, Prevention of Environmental Pollution, WHO, Geneva, Switzerland (*Secretary*)